# STEAM & Me™

# SPACE EXPLORATION

## L. J. TRACOSAS

Starry Forest Books

# STEAM & Me™

**SCIENCE · TECHNOLOGY · ENGINEERING · ARTS · MATHEMATICS**

Draw a super-smart robot. Create your own wind energy. Find out if your teeth are as sharp as a shark's. Go back in time to the world of dinosaurs or rocket into space. Power up that scientific brain of yours with **STEAM & Me**!

Photos, facts, and fun hands-on activities fill every book. Explore and expand your world with science, technology, engineering, arts, and math.

# STEAM&Me is all about you!

**Great photos** to help you get the picture

**New ideas** sure to change how you see your world

There are two types of planets in our solar system: gas planets and terrestrial planets.

## Other planets orbit the sun, too.

A *solar system* is a star and the planets that orbit around it. Our solar system has eight planets. Closest to the sun is Mercury, followed by Venus. Next is Earth—that's us! Then there's Mars, Jupiter, Saturn, Uranus, and Neptune.

Which planet in the solar system is **Earth**? Point to it.

### Gas Giants
These planets are made of gases. The gas planets are Jupiter, Saturn, Uranus, and Neptune.

### Terrestrial Planets
These planets have hard surfaces made of rocks and minerals. The terrestrial planets are Venus, Mercury, Earth, and Mars.

**STEAM&Me**

Look at the picture of the solar system on this page. Which planet is the biggest in our solar system? Which planet is the smallest? One way to remember the order of the planets is to make up a sentence using the first letters of their names, such as: **M**y **V**ery **E**xcellent **M**other **J**ust **S**erved **U**s **N**achos. Make up your own sentence to help you remember the order of the planets.

**Fascinating facts** to fill and thrill your brain

**Hands-on activities** to spark your imagination

# Take off into space!

Do you ever look up at the night sky and wonder what it's like up there with the stars? People have been looking at the stars for thousands of years. But we've only just started to travel into space and learn more about our solar system and the universe. There's so much to explore! Let's blast off and see what's out there.

It's out of this world! In boxes like these throughout the book, you'll think about your place in the universe, what it would be like to live in space, and more!

STEAM & Me

# Shapes in the Sky

**Constellations** are groups of stars that form shapes or patterns. Can you see the stars that form the Big Dipper?

Look down there! It's planet Earth. Astronauts took this photo from space.

## Where Is Space?

Space starts about 60 miles above Earth. That's more than 1,000 football fields away.

# What is space?

Look up. Past the blue sky is a place where there is no air and no light. There's no sound, either. That place is space.

## Your Flight Is Canceled!

You can't take an airplane to space. Jets fly high in the sky. But jet engines need air to work. Because there's no air in space, an airplane can't fly there.

# There's no light and no air. But space isn't empty.

Space is filled with many things. It has small things like gases and *particles*, or teeny-tiny pieces of matter. Space also has big things like planets, moons, and stars.

Look up into the night sky. What shape is the moon? Can you see the Big Dipper? Invent your own constellation; use your fingers to connect the stars you see. Make up a cool name for your constellation.

**STEAM & Me**

# Space Dust

*Asteroids* are rocks that orbit in space. Space dust can form when objects like asteroids crash into moons or other asteroids.

It takes Earth 365 days, or one year, to travel around the sun.

## Rotating the Day Away

As Earth spins, or rotates, we get daytime and nighttime. Daytime happens in the places on Earth facing the sun. Nighttime happens when a spot on Earth faces away from the sun.

# A planet is a round body that orbits around a star.

You're on a planet right now. Your planet's name is Earth. Earth is circling around, or orbiting, a star. That star is the sun. Earth is also rotating, or spinning. Are you dizzy yet?

## Zoom!

Earth is speeding around the sun at 67,000 miles per hour. That's about 800 times faster than a race car goes around a track.

**STEAM & Me**

Each year, you take a trip all the way around the sun. How old are you? How many full orbits have you made? How long will it be until your birthday, when you'll start your next orbit?

# The star that Earth orbits is the sun.

The sun is a star just like the stars you see twinkling in the night sky. The sun looks different because it is so much closer to Earth than any other star. Like all stars, the sun is a big ball of hot gases. It's the heat from the gases that makes the sun bright.

Light and heat from the sun travel to Earth in about **eight minutes**. That light and warmth make life on Earth possible. Nothing—including you—can live or grow without the sun.

## Mass of Gas

The sun is made of gases called *hydrogen* and *helium*. Helium is the same stuff that makes balloons float.

## One Big Star

The sun is so big that more than a million Earths could fit inside of it.

There are two types of planets in our solar system: gas planets and terrestrial planets.

## Gas Giants

These planets are made of gases. The gas planets are Jupiter, Saturn, Uranus, and Neptune.

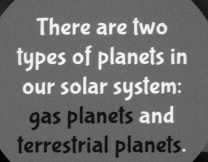

## Terrestrial Planets

These planets have hard surfaces made of rocks and minerals. The terrestrial planets are Venus, Mercury, Earth, and Mars.

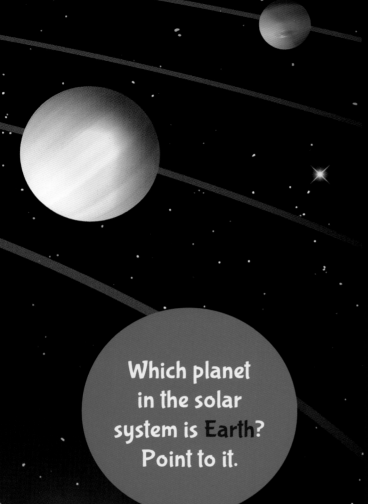

# Other planets orbit the sun, too.

A *solar system* is a star and the planets that orbit around it. Our solar system has eight planets. Closest to the sun is Mercury, followed by Venus. Next is Earth—that's us! Then there's Mars, Jupiter, Saturn, Uranus, and Neptune.

Which planet in the solar system is Earth? Point to it.

**STEAM & Me**

Look at the picture of the solar system on this page. Which planet is the biggest in our solar system? Which planet is the smallest? One way to remember the order of the planets is to make up a sentence using the first letters of their names, such as: **M**y **V**ery **E**xcellent **M**other **J**ust **S**erved **U**s **N**achos. Make up your own sentence to help you remember the order of the planets.

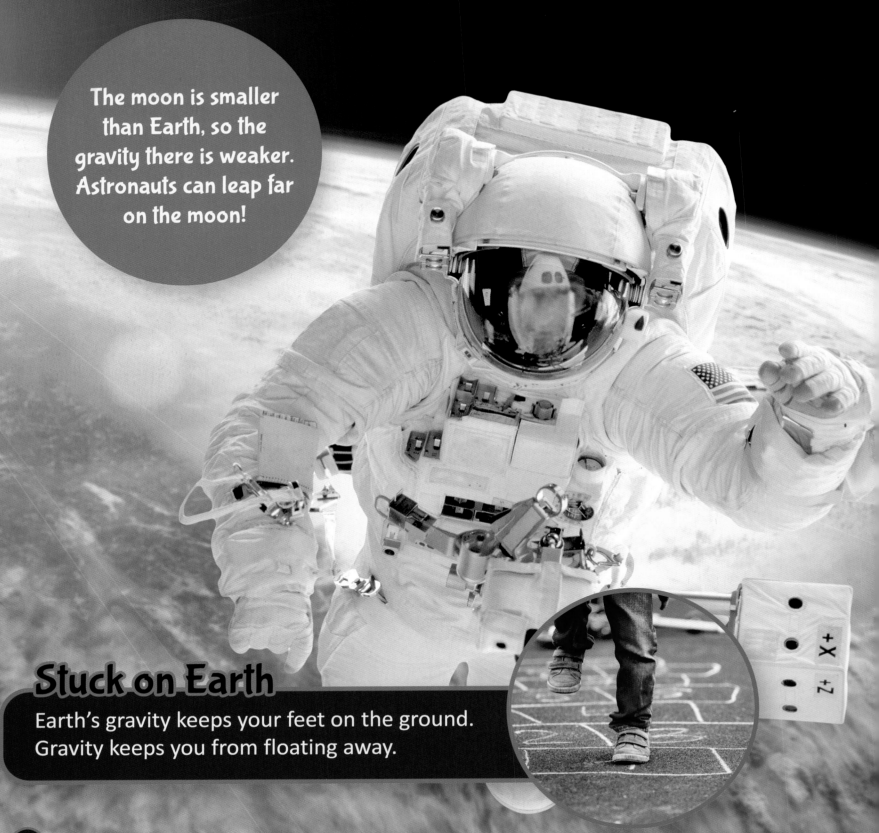

The moon is smaller than Earth, so the gravity there is weaker. Astronauts can leap far on the moon!

## Stuck on Earth

Earth's gravity keeps your feet on the ground. Gravity keeps you from floating away.

that pulls objects toward each other. The sun is so big that the strength of its gravity keeps the planets from spinning away into space.

## Weight Way Out There

*Mass* is a measure of how much matter something is made up of. The larger mass something has, the more gravity it has. Jupiter has a larger mass than Earth, so Jupiter has stronger gravity. The strength of gravity also affects how much things weigh. A cat weighs about 8 pounds on Earth. But on super-large Jupiter, that same cat weighs about 20 pounds!

# Is the moon a planet?

The moon might look like a planet because it's round. But planets orbit stars, and the moon orbits Earth, so it is not a planet. The moon is not a star, either, even though it shines like one. The moon is really a big rock. It glows only because light from the sun shines on it.

## Now You See It, Now You Don't

Sometimes there's a bright full moon, big and round in the night sky. Other times, just half or a sliver of the moon shines at night. Sometimes we can't see the moon shining at all—but the whole moon is up there all the time. As the moon orbits Earth, we get to see different parts of it every night.

Look at the moon every night for a week and draw its shape as it changes. Is the moon getting bigger—*waxing*—or is it getting smaller—*waning*?

The moon's surface is covered in craters, or big dents. The craters form when asteroids, or space rocks, hit its surface.

## Many Moons

Earth has just one moon. Other planets in our solar system have many more. Jupiter has more than 70 moons, and Saturn has more than 50 moons.

## Are We There Yet?

Proxima Centauri, our nearest star neighbor, is about 25 trillion miles away. It would take thousands of years to fly there in a space shuttle.

People haven't yet invented a spacecraft that can fly to distant stars. Use your imagination and draw a picture of the spacecraft you would create to fly to Proxima Centauri. Would it be shaped like a rocket, or something else completely? How big would its rocket boosters be?

# You can see about 2,500 stars at night.

With special equipment, you can see millions of stars. Stars are very far away—so far that we don't measure the distance in miles. We measure the distance to stars in *light-years*. One light-year equals trillions of miles.

Very powerful equipment allows scientists to peer far into space. The Hubble Telescope took this photo showing millions of stars.

# Have you ever seen a shooting star?

If you have, it wasn't a star at all. It was probably a **meteor**—the flash from a space rock called a meteoroid as it burned up in Earth's atmosphere. A meteoroid is just one type of object that zooms through space.

Look! There's a meteor.

## Asteroid

*Asteroids* are rocks that zoom around the solar system.

22

## Comet

*Comets* are balls of ice and dust that come from outside the solar system.

## Meteoroid

*Meteoroids* are space rocks that are bigger than a speck of dust and smaller than an asteroid.

## Meteorite

A *meteorite* is a space rock that lands on Earth.

## Is There Milk in the Milky Way?

No! The Milky Way gets its name from the bands of dust and gas that swirl in it. The clouds look a bit like milk. On a dark night, you can sometimes see those milky clouds from Earth.

Galaxies, or collections of stars, can be shaped like spirals. When a very powerful **telescope** called the Hubble Telescope looked up in the night sky, it found more than 10,000 galaxies in just one spot of space.

# Planet Earth is in the Milky Way galaxy.

A **galaxy** is a collection of stars and their solar systems, plus swirling gases and dust. The Milky Way is a galaxy in the shape of a spiral. Gravity holds a galaxy together.

## What Else Is Out There?

You can see light from stars and galaxies in space. But there's stuff there we can't see, too. A *black hole* is an area of space that has so much gravity that not even light can escape it.

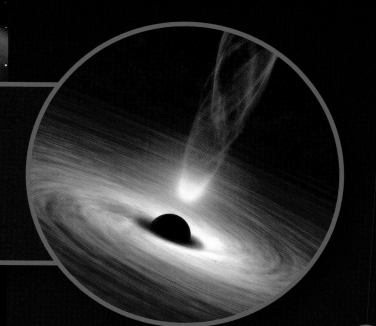

# Up, up, and away!

People who are trained to travel to space are called *astronauts*. Scientists, engineers, pilots, mathematicians, and other experts can study to become astronauts and learn how to travel and live in space.

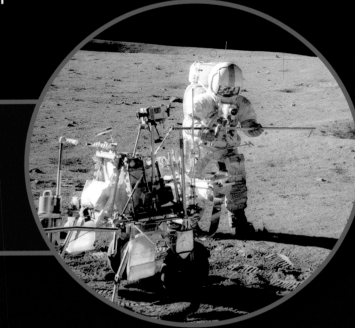

## Fore!

Astronaut Alan Shepard, commander of the Apollo 14 space mission, played golf on the moon. He hit two golf balls— and they're still up there on the moon!

Would you like to live in space? What would you do there? How do you think it would be different from living on Earth? Draw a comic book page that shows what your life might be like in space.

STEAM & Me

Without Earth's gravity to hold them down, astronauts are weightless in space— and so is their food!

27

The space shuttle *Atlantis* takes off from the Kennedy Space Center in Florida. The rocket booster blasts the shuttle into space and then falls away.

Rockets burn fuel, which creates the energy needed to blast off into space. Once a rocket boosts a shuttle high enough, the rocket falls away. It floats to Earth on a parachute. In space, people need a space shuttle to get around. Space shuttles are designed to move in space, and to come back down to Earth.

## Living in Space

Fifteen countries worked together to build the International Space Station. That's a lab where astronauts live and study space. People are living there right now!

# There's so much space to explore.

From faraway planets to even farther away galaxies, there's lots to discover in space. Imagine a space shuttle is taking off, and you are riding in it. Where do you want to go? Draw a map of your route and make a list of what you would take with you.

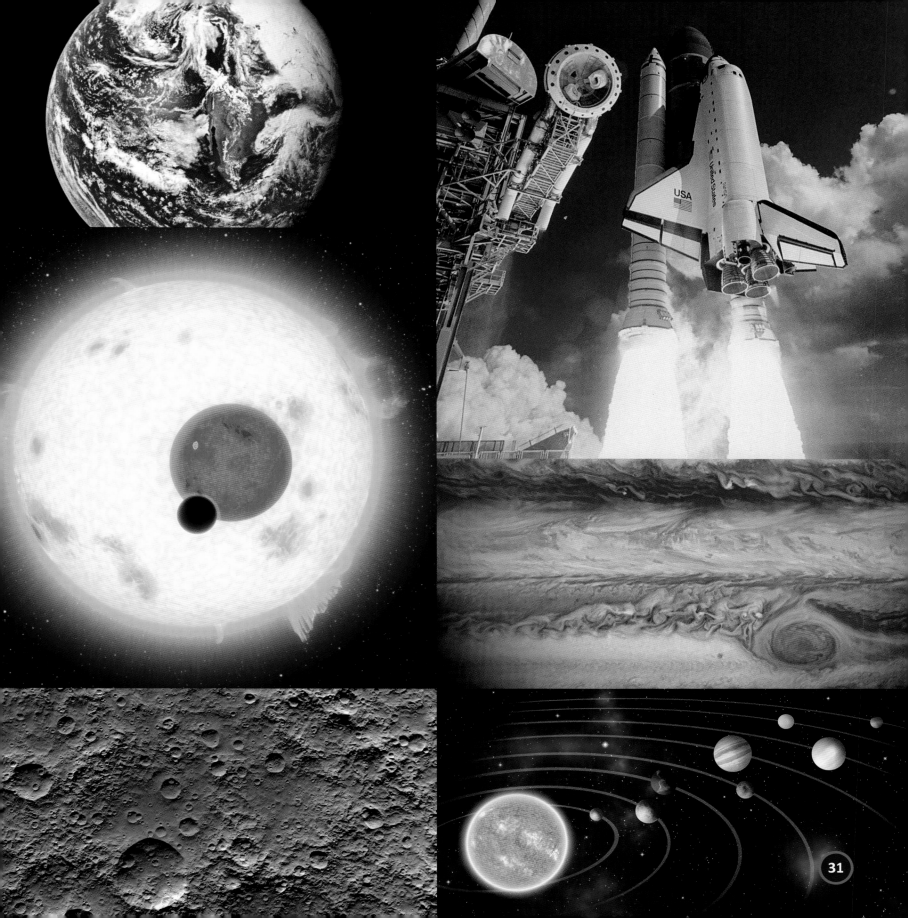

# Glossary

*Learn these key words and make them your own!*

**constellation:** a named group of stars in the sky. *The stars in the Big Dipper* constellation *form the shape of a big square cup.*

**galaxy:** a large group of stars in the universe. *Earth is in the Milky Way* galaxy.

**gravity:** a force that draws objects together. *Our planet's* gravity *keeps us from floating away.*

**meteor:** the flash of light from a meteoroid that enters Earth's atmosphere. *The* meteor *was a bright flash in the sky.*

**satellite:** a human-made machine sent into space to orbit a planet or an object. *The* satellite *orbits Earth and sends data to scientists.*

**telescope:** a tube-shaped tool with lenses to see faraway objects. *I saw craters on the moon when I looked through the* telescope.

---

*For Miles, my star in the sky.*

STEAM & Me and Starry Forest® are trademarks or registered trademarks of Starry Forest Books, Inc. • Text and Illustrations © 2020 and 2021 by Starry Forest Books, Inc. • This 2021 edition published by Starry Forest Books, Inc. • P.O. Box 1797, 217 East 70th Street, New York, NY 10021 • All rights reserved. No part of this publication may be reproduced, stored in a retrieval system, or transmitted in any form or by any means (including electronic, mechanical, photocopying, recording, or otherwise) without prior written permission from the publisher. • ISBN 978-1-946260-90-1 • Manufactured in China • Lot #: 2 4 6 8 10 9 7 5 3 1 • 03/21

Photo credits: ASP: Alamy Stock Photo; SS: Shutterstock. Cover, Vladi333/SS; 4-5, ANON MUENPROM/SS; 5, Olha Polishchuk/SS; 6, NASA; 6-7, NASA; 7, Frank_peters/SS; 8, Dotted Yeti/SS; 9, NASA/JPL-Caltech; 9, (LO) NASA/JPL-Caltech/S. Stolovy (SSC/Caltech); 10, JBArt/SS; 10, (LO) XiXinXing/SS; 11, Abdul Razak Latif/SS; 12, Sunny studio/SS; 12-13, Aphelleon/SS; 13, Yuliia Markova/SS; 14, (LO LE) NASA images/SS; 14, (LO RT) NASA images/SS; 14-15, Withan Tor/SS; 16, Juergen Faelchle/SS; 16, (LO) NadyaEugene/SS; 17, Lightspruch/SS; 18, Designua/SS; 19, HelenField/SS; 19, (LO) Mirai/SS; 20, Vadim Sadovski/SS; 20-21, ESA/Hubble & NASA; 22, Stephane Masclaux/SS; 22-23, Eliyahu Ungar-Sargon/SS; 23, (UP) ESO/E. Slawik; 23, (CTR) dotted zebra/ASP; 23, (LO) Rostasedlacek/SS; 24, Evelyn Meis/SS; 24-25, Alex Mit/SS; 25, NASA/JPL-Caltech; 26, NASA/Edgar Mitchell; 27, NASA; 28, NASA/KSC; 29, NASA; 30, (Stars) ESA/Hubble & NASA; 30, (Milky Way) Alex Mit/SS; 30, (Black Hole) NASA/JPL-Caltech; 30, (Asteroid) Stephane Masclaux/SS; 31, (Earth) NASA; 31, (Kepler 16b) NASA/JPL-Caltech; 31, (Moon) HelenField/SS; 31, (Shuttle) NASA/KSC; 31, (Jupiter) NASA images/SS; 31, (Solar System) Withan Tor/SS; 32, (CTR) NASA; 32, (LO) Trevor Hales/SS; Back cover, (UP) Yuliia Markova/SS, (LO LE) NASA/KSC, (LO CTR) Stephane Masclaux/SS